Robert Stevenson

A New Potential Principle in Nature

Robert Stevenson

A New Potential Principle in Nature

ISBN/EAN: 9783744674577

Printed in Europe, USA, Canada, Australia, Japan

Cover: Foto ©berggeist007 / pixelio.de

More available books at **www.hansebooks.com**

ELASTICITY A MODE OF MOTION

BEING A POPULAR DESCRIPTION OF

A NEW AND IMPORTANT DISCOVERY
IN SCIENCE

—— BY ——

ROBERT STEVENSON, C.E., M.E.

Member of the American Institute of Mining Engineers; California Academy
of Sciences; Technical Society of the Pacific Coast, etc., etc.

April 20, 1895.

Dear Sir:

I beg to forward per book post a pamphlet containing a short explanatory essay on my "Discovery of the TRUE and PROXIMATE Cause of Universal Gravitation."

I have tried to make the description so plain that at a glance you will easily perceive the TRUTH of the discovery as well as its importance to Science.

Being a non=professional and silent investigator, in scientific subjects, my name and influence cannot be expected to give this matter the right to professional consideration which it deserves.

I therefore desire you to assist in a critical discussion of the subject in any way which seems to you most suited to arouse professional interest thereon.

Yours very respectfully,

Robt. Stevenson, C.E.,

Member of the Academy of Sciences.

P. O. Box 2214,
San Francisco, Cal.

PREFACE.

The discovery which I am about to describe was first impressed on my attention twenty-eight years ago, when a student attending Sir William Thomson's class in the old college of Glasgow University. But I did not thoroughly appreciate the importance of the matter until about fifteen years ago. Since then, although cognizant of the true cause of gravitation, I have not been in a position to take advantage of the immense power, and wonderful insight, that such knowledge confers.

My business as a mining engineer requiring all my time and attention, and not being a recognized authority in natural sciences, my desultory efforts from time to time to call the attention of the recognized authorities on the subject, to the facts of the case, have met with no success.

I am now preparing to publish my notes, containing the mathematical proofs and experiments by which I have established the truth (to my own satisfaction) of the great principle in Nature which is an intermediate and universal agent between the spiritual agency of force, and the material energy which produces gravitation and evolution.

I have so far been able to trace this intermediary principle, as the great potential, and conservative energy in the material universe. Its elastic resistance to the radical forces of dispersion, being as necessary to the stable existence of a molecule as it is to that of the planetary systems of the whole universe.

ROBERT STEVENSON, C. E.

Academy of Sciences, San Francisco.
April 10, 1895.

CONTENTS.

6 CONTENTS.

ELASTICITY OF MOTION.

THE TRUE AND PROXIMATE CAUSE OF UNIVERSAL GRAVITATION.

(1) It may appear to be egotistic in me to say that the principle of *Kinetic Stability*, or that which produces it, which I call *The Persistence of Energy*, is the greatest natural principle that science has ever discovered, and when it becomes thoroughly known, it will lead to a harvest of knowledge and power for mankind, such as the world has never before experienced.

(2) It will in fact prove the heavenly key to a realm of spiritual truth, whose grandeur consists in this, that it will enable mankind to understand Nature better, and to utilize her more intelligently for the benefit of the human race and the glory of the Divine Individuality, who has created by the spirit of His power all the phenomena of the universe.

(3) It would be out of place in this short explanatory paper to enter into the historical portion of the subject of gravitation; nor do I propose to discuss the merits of the various theories which have been advanced during the last three hundred years.

(4) I propose therefore to devote myself to the exclusive task of explaining the theory and its application, which I am about to publish, so that those naturalists who take an interest in the great physical principles may understand before hand, and be prepared to criticise the work when it appears.

(5) Before entering on the explanatory portion of the subject, allow me to say that to the illustrious Newton belongs the great and undying honor of having rescued the theory of gravity from the grasp of metaphysics, and established it on the firm basis of physical science, where for two centuries it has resisted every attempt to displace it.

(6) And if it should be the duty of science now to acknowledge that Newton was mistaken, both in his hypothesis and his theory, still the

credit of having discovered and demonstrated the law of the inverse square of the distance, by means of which astronomy has made such great progress, will ever remain as a halo of glory around the illustrious author of the "Principia."

(7) According to the accepted laws of motion, curvilinear motion is impossible without a centripetal force.

(8) Newton started on that supposition, and with Kepler's laws as mathematical data, he demonstrated the great law of universal gravitation, on which all theoretical astronomy is based, and from the application of which many great discoveries have been made in the times, distances, and motions of the planets and comets of the solar system.

(9) But if it can be shown that the hypothesis of a centripetal force is unnecessary, then I think there will be no difficulty in establishing the fact that orbital motion in free space is not caused by centripetal force.

(10) In our every-day experience it is very seldom that we observe a case of free motion in

free space. The nearest approach to it that we meet with is in the case of the cricketer, and baseball player, and curious to say, both of them can throw a "curved ball," a ball which on leaving the hand of the thrower is made to describe a horizontal curvilinear path; and I can prove that if it were not for the resistance of the air, and the force of gravity, it would be possible to make a ball describe a complete horizontal circuit round an imaginary point as a center without any tension between the center and ball.

(11) By this hypothesis the ball receives a longitudinal and transverse energy in the same plane impressed upon it, which causes it to describe a curvilinear path, or in other words, the resultant of two transverse rectilinear energies given to a free body in free space is curvilinear motion.

(12) What we have been taught of the principles of the composition and resolution of forces; motions, energies, etc., in Dynamical Science, will predispose us to criticise that proposition, but if we will consider for a moment the different conditions between a free body in free space and a body which is constrained to move in a fixed

direction, we will see that the principle which applies to the one may not apply to the other.

(13) As I have said, this paper is only explanatory of the principles on the proof of which the new theory of gravitation rests.

(14) To prove all the propositions necessary to establish these principles requires a good-sized volume, which I hope to have ready for publication in a very short time, as the manuscript is nearly completed now.

(15) To thoroughly understand the theory, astronomers must be prepared to look at the subject from a different point of view than that from which they have been taught and accustomed to, as the theory bears a similar relation to the Newtonian theory, that the Copernican bears to the Ptolemaic, and in both cases it is simply the difference between an apparent action and a real action.

(16) In the following diagram let us suppose the mass m to be a body moving in the straight line A B, and for simplicity's sake let us con-

sider it as a dynamical particle, moving with uniform velocity v in the direction of B with independent motion in free space.

FIG. 1.

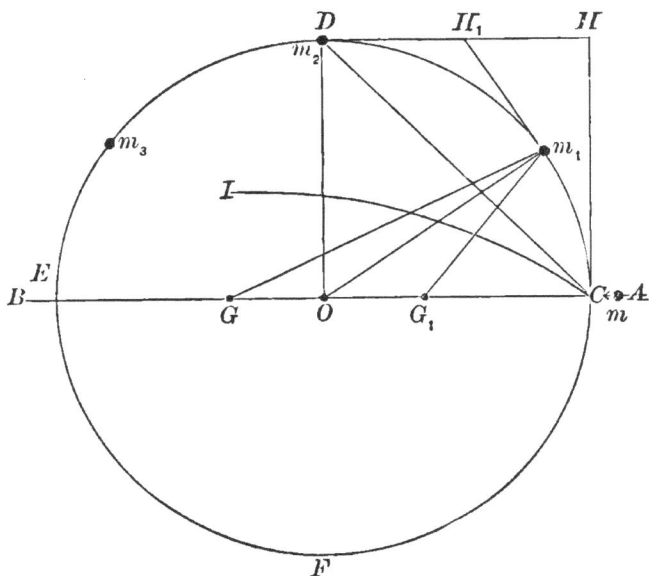

Then the kinetic energy of $m = \dfrac{m \, v^2}{2} = E_k$.

(17) When m reaches the point C, let us suppose that a transverse impulse superimposes on m an additional kinetic energy E_k in the direction of H along $C H$. Now m is moving with twice the kinetic energy, and therefore its resultant velocity is represented in direction and

magnitude by the straight line $C\,D$, which is the diagonal of the parallelogram $C\,O\,D\,H$, that is what is taught in the universities at present, and that is one point where my discovery proves that such teaching is wrong.

(18) I find by experiment, and I can also prove it by physical geometry, that m when moving with kinetic energy along $A\,B$ has also *kinetic stability*, due to what I call its *persistence of energy*, and that the kinetic stability acts as a latent or potential energy, which acts with reciprocating effect against a deflecting force or impulse along $C\,H$, and consequently the resultant, instead of being along the diagonal $C\,D$, is along the curvilinear line $C\,I$, and we can show that the path is elliptical.

(19) We also prove that the kinetic stability $= E_s$ of m along $A\,B$ is equal to the kinetic energy, and that $E_k + E_s = E =$ total energy, or *vis viva*, and that E_s the kinetic stability is inversely proportionate to E_p the potential energy.

(20) By these propositions we prove that the motion of a free body, such as a planet, is elastic,

and that its orbit is due to the superimposed motions, hence, although it appears to be attracted to a focus, it really contains within itself a couple and a direct force, which constrains it to move in its orbit without any attractive or repulsive force, or any elastic medium of any kind whatever. It is in itself endowed with elastic motion.

(21) I can also prove that no force, however great, unless it be infinite, can ever permanently deflect a free moving body from the line of its previous motion.

(22) Let m be acted on at C with an impulse as great as you please, which is not infinite, and there is always a component tending to bring it back to the line $A\ B$, and such a component is inversely proportionate to the radius of curvature at any point in the body's motion.

(23) Now this fact explains the stability of the solar system. Supposing the sun had kinetic energy due to its inertia and longitudinal velocity, then by some means a transverse impulse is superimposed at a radial distance from its center of inertia, acting as a couple with the

kinetic stability, producing rotation as well as linear velocity, then I can prove that the whole energy due to rotation would be absorbed into the sun's mass as heat; and if it was great enough the mass would become incandescent, and sufficient material might be dispersed as incandescent vapor to produce a nebulous atmosphere, stretching to the very confines of the present solar system, out of which, as Laplace has shown, the whole planetary system could be evolved by refrigeration, and the persistence of energy of the molecular vortices within the nebula.

(24) In such a system as this, we can prove that the tendency of each molecule to return to its original line of motion (its persistence of energy) would always cause its apparent centripetal energy to more than balance its centrifugal energy. Hence the *Force of Restitution*.

(25) Returning again to the diagram, Fig. 1, let us suppose the mass m_1 m_2 m_3 to be a body moving in the orbit $C D E$, which we will suppose to be a semi-ellipse, whose center is O, and the foci G and G_1, and $C E$ the major axis or diameter, which we may extend both ways to A

and B, then suppose m_1 to move from C to D with a uniform and independent motion.

(26) Then according to the usual interpretation of the Newtonian laws of motion, for m_1 to describe the curve $C\, m_1\, D$ it is necessary that it should be acted on by a force towards H in the tangent $C\, H$, and another and independent force in the direction $C\, E$, and that the force along $C\, E$ should always be at right angles to the tangent at any point of the curve $C\, m_1\, D$, so that when the body m reaches m_1 the direction of the composite forces would be along the tangent to the curve at m_1, and the radius vector $m_1\, G$ if the orbit was elliptical, and $O\, m_1$ if circular.

(27) Now as a purely geometrical, or kinematical problem, that way of looking at it is correct, but when we have to deal with a body which has the physical property of kinetic energy, due to its independent motion, we have to adopt other methods than that used by Newton in the "Principia" to correctly solve the problem.

(28) If we suppose $O\, m_1$ to be a string fixed at O as a pivot, and attached to m_1, or if we sup-

pose the string $G\,m_1\,G_1$ fixed at the foci G and G_1 passing through the center of m_1, then we know that such an arrangement will produce circular or elliptical motion if m_1 be acted on by an independent tangential force, such as that along $C\,H$ or $m_1\,H_1$, and the tension on the string would vary constantly with the variation of curvature, and be proportional to the curvature.

(29) Now instead of the string, the Newtonian theory requires an attractive force at one of the foci, say G, acting as a tension between G and m_1, which varies as the multiple of the masses at G and m_1 divided by the square of the distance between their centers of gravity, which is:

$$\frac{G \times m_1}{G\,m_1^{\,2}} = \frac{m \times m_1}{D^2}$$

is the Newtonian expression for the universal law of gravitation. A very simple law when we know the true value of each respective mass, and a law which must always be correct when the inverse square of the distance is used to calculate the value of the mass.

(30) It is not unscientific to suppose that a force may decrease as the square of the distance increases, on the other hand, it is not only unscientific, but it is perfect nonsense to say that

the resultant force of attraction between two masses varies as the multiple of the masses, when we know that the natural laws only require addition instead of multiplication.

(31) But, as I say, for pure kinematical motions such a law $\frac{m \times m_1}{D^2}$ is absolutely correct when the value of the mass is found by the inverse square of the distance.

(32) Now whilst such a law may give us the true relative motions of the planets and comets of the solar system, it can never give us the true specific mass or gravity, or the specific energy of each body separately, hence the limit to the utility of the Newtonian law. The law is indeed a kinematical law, whereas the new law will be kinetic, and therefore become the great law of energy, not only in the inorganic but also in the organic and spiritual worlds, and on the proper application of this law not only does the conservation of natural energy depend, but the truth and value of eternal life becomes demonstrable.

(33) In the case of the solar system, with the sun at G, and the planet at $m_1 \, m_2 \, m_3$, the centripetal force looks as if it might be attractive, but

if there had been no sun in the focal center of the solar system, and only empty space, as we will show might happen in Nature, then the theory of attraction could never have existed.

(34) The supposed action at a distance we will prove to be only apparent, just as Ptolemy's supposed motion of the stars round the earth was apparent only.   It is therefore quite unnecessary that there should be a "plenum of vortices," or an "elastic semi-material medium," filling interplanetary space to produce attraction or apparent tension.

(35) My theory is based on the discovery that a mass m moving with uniform or accelerated velocity in the line A B, with independent motion in empty space, may have another independent motion superimposed on it at C, such that its resultant path (due to these two motions alone) would lie along the semi-ellipse C D E; or in other words, that the resultant of two transverse independent free motions, which have kinetic energy in free space, lies along a curve instead of the diagonal of the parallelogram, of which the adjacent sides represent the composite motions.

(36) The experimental proof of this fact is very simple and convincing. Take a magnetized steel ball two inches in diameter, project it from a spring catapult with a velocity of 50 feet per second along a perfectly flat and smooth table, 20 feet long by 6 feet wide, dust the top of the table with a white powder, and cover the ball with lamp black, then at a distance of one 'foot from where the ball leaves the catapult spring have a fixed electro magnet so arranged as to give an instantaneous transverse repulsion to the ball equal in energy to what the ball received from the catapult, and the line of resultant motion will be seen to be elliptical. If the experiment was carried out in a vacuum, and the surface of the table and ball were perfectly true and smooth, the path would be perfectly elliptical.

(37) At one time I thought the proof could be better established by using a vacuum tube in some tall tower, such as the Metropolitan Tower, but Lord Kelvin, to whom I mentioned the matter, thought it would not answer, and after considering the matter I saw that he was right, because the tower being a portion of the earth, a body falling within the tower would be moving under constraint, and the resultant

motion in that case would lie along the diagonal of a parallelogram.

(38) The rule is: That when an independent motion is transversely superimposed on a constrained motion the resultant lies along the diagonal straight line, but when an independent motion is superimposed on another independent free motion, the resultant is along a continuous curve.

(39) For instance, if a man walks across a railway carriage whilst it is moving in a straight line, the resultant due to the composition of the two motions is the diagonal straight line, according to the well-known principles of dynamics.

(40) On the other hand, if the man was to jump from the side of the moving carriage, and was not acted on by any other forces than those due to his own kinetic energy (the one parallel to the carriage, and the other transverse), then he would describe an elliptical curve, and return to the point he jumped from, if the motion of the carriage was uniform.

(41) The difference in these two cases depends on the different degrees of freedom of motion. In the first case, whilst walking across the moving carriage, the mass of the body had momentum in the direction of motion of the carriage, and kinetic energy in a transverse direction; whereas in the second case the body had not only kinetic energy transverse, but also kinetic energy parallel to the motion of the carriage, and if the mathematicians will compound these two co-ordinates they will find at once that the resultant must be along a curve whose equation may be expressed thus:

$$\frac{x^2}{x_1{}^2} + \frac{y^2}{y_1{}^2} = 1$$

when x and y are the co-ordinates of the body, and x_1 and y_1 the semi amplitudes.

(42) When we consider the difference between momentum and kinetic energy it will be found that the mathematical proof of the theory is not very difficult, and by adopting a method of physical geometry the proof becomes not only convincing, but absolutely correct and incontrovertible.

(43) We prove thus by means of a number of propositions that free motion in free space is elas-

tic, and follows absolutely the laws of elasticity.

(44) That the same laws which govern the Stellar phenomena control and produce the atomic motions, and regulate their affinities so as to produce molecular cohesion.

(45) That light, as the earliest and least material energy in the universe, is due to the vibratory action of the spirit of power, or according to science it is due to the elastic vibration of radiating atoms, or lines of force in a non-material field of force.

(46) That heat is caused by inter-molecular vibration.

(47) And electricity is produced by the interference of waves of molecular disruption.

(48) That the sun's heat is due to incandescence produced by the inter-molecular vibrations of the superimposed kinetic energies, due to the sun's straight and rotary motions, and that fact can be proven by the new law with the help of Joule's equivalent.

(49) It can also be proven by this theory that the continuous horizontal velocity of a free body in a vacuum near the earth's surface at 1,036 feet per second will prevent that mass from falling to the earth; or, in other words, that the kinetic stability of a body moving with independent kinetic energy transverse to the force of gravity may tend to overcome gravity to such an extent that the body may move in a circle concentric to that of the earth's circumference.

(50) These, and a number of other facts regarding the motions of the solar system, especially the specific energy inherent in the mass of each planet, and their respective inductive capacities, both for light, heat and electricity; also an explanation of the cause of the tides according to the new theory, will be given more fully in the book I am about to publish on the subject.

(51) In conclusion I feel bound to support the prophetic saying of that great seer the Abbe Moigno, who in the *Cosmos* of 1852 declares:

"That if there is anything certain in this "world it is that the molecules of bodies and "bodies themselves are not really self attractive; "it is that attraction is not an intrinsic but only

"a developed force; it is that notwithstanding
"everything occurs as though bodies mutually
"attracted each other; it is incontestably true
"that bodies do not attract. Newton, as Euler,
"as every philosopher worthy of the name, has
"seen in Nature but two things, *inertia* and
"*motion*, originally impressed by a free will, the
"the first and infinite mover. And it is with
"these two great facts of inertia and movement
"that advancing science shall ultimately explain
"all the phenomena of the physical world."

THE NEW KINETIC THEORY OF GRAVITATION.

[A paper read before the Technical Society of the Pacific Coast.]

PART I.

The old established theory of centripetal force is that given by Sir Isaac Newton nearly three centuries ago, in the memorable words of the " Principia ": " That every particle of matter in the universe attracts every other particle," and that every deviation from a straight line in absolute free motion is due to that fact of mutual attraction.

That is the force which bends the body's motion and keeps it revolving in an orbit.

Newton himself in the "Principia " compares it to the string of a sling which keeps the stone from flying away from the centre.

Whilst experimenting with bodies at high velocities, I found that it was not necessary to have a solid centre of attraction to produce this centripetal force. I found that bodies in motion are apparantly attracted to imaginary points, lines and planes, where no attractive force can possibly exist; and during the last fifteen years have given this fact some study in my spare hours, and have discovered what I think is the true cause of this centripetal force, and am now prepared to put it forward in a definite form for the acceptance of science.

I am fully aware that deep rooted and established ideas are difficult to remove in the scientific, as well as in the religious world.

We all know how long it took the scientific world to comprehend the idea that it was the weight of the air which forced the water to rise in a vacuum, the " Fuga Vacui " having been the accepted cause for thousands of years.

We also know how successfully the great mathematician Ptolemy held the scientific world in mental bondage by his beautiful theory of the earth as the centre, round which the whole heavens revolved every twenty-four hours.

Now I find that the illustrious Newton has fallen into a similar delusion, by means of the analogy of the sling and stone he failed to comprehend the possibility of constrained motion in an orbit without a pull towards the centre or the focus of the orbit. At first sight it would appear to an observer that there must be a pull to divert a moving body from the line of its motion, indeed the first and second laws of motion in dynamics clearly show that there must be such a force. And Newton was quite justified

in calling it a centripetal force. I accept his
definition of a centripetal force, and I accept his
law of the inverse square of the distance, as a
true but empirical law of force, but I disagree
with his theory that this centripetal force is in its
nature or action "attractive."

This false theory of attraction has done much
to obstruct the progress of physical science; to it
is due the fact that mathematical analysis is
almost powerless in dealing with high speed
machines, and fluid motion at high velocities, and
that all the formulæ in the molecular constitution
of bodies and chemical physics, as well as the
motions of the heavenly bodies, are nearly all
empirical. The new theory, which I call the
Theory of the Persistence of Energy, will when
properly understood open up to mathematics a
right of way by which all these branches of
science will be fully explored.

As this paper is only intended to be pre-
liminary, and to give just sufficient information
as to lead to the discussion of the subject, I will
confine myself to the first principles of the subject where they diverge from the established
theory of attraction, and leave the application of
the theory until we have satisfied ourselves that
the theory itself is correct.

As you may be aware I have already delivered

a lecture at the Academy of Sciences, on the historical portion of the theory of gravitation, and there discussed the Newtonian theory of attraction; quoting Newton's own words to show that he never believed in the innate and inherent force of attraction in matter, but he always hoped to be able to prove that attraction was caused by the motion of an aetherial medium. All the great natural philosophers since his day have worked at that problem, either to prove that apparent attraction was due to the vibrations of an aetherial medium, or to the vortices of radiant atoms.

Faraday came nearest to an acceptable proof when he discovered that magnetic attraction was due to the inductive action of a field of force, which surrounds the poles of the magnet.

Now, you see the question of the attraction of matter is by no means a settled question, although the editor of our only scientific paper, in discussing my lecture before the Academy of Sciences, thought I was trying to destroy a law of nature; and consequently made some silly remarks about me, which in the daily press would have been very amusing, but from a scientific journal and the only scientific journal of the Pacific Coast, makes one wish for a scientific journal which understands science.

Now I defy any scientist of standing to contra-
dict me in the assertion that science has never
definitely adopted action at a distance (as
attraction is called) as a scientific truth. It has
been accepted as a scientific dogma awaiting the
true solution of the cause of attraction, hence it
is quite legitimate on the part of any scientist to
try to solve the problem, and he who can discover the true cause of gravity has no need
to be ashamed of his discovery.

It is only a few months ago since the greatest
of all the great natural philosophers of this age,
Lord Kelvin, stated to an admiring audience of
scientists in London, that this great problem had
for fifty years been constantly in his mind,
sleeping and waking it was always there, and
yet we are told by our only scientific press that
any professor who would spend his time in
trying to discover the cause of gravity had most
assuredly mistaken his profession. Well gentle-
men, I hope before we are through with the
discussion of this subject there will be no doubt
in your minds as to which has mistaken his
profession, the editor of the scientific press, or
your humble servant.

In my lecture before the Academy, I gave six
illustrations of motion, which in each case
showed the impossibility of mutual attraction

between matter, but I will in this paper confine myself to one crucial test of the attraction of the earth to matter on or near its surface; a test which this Society can make at very little expense.

We all know that the law of the attraction of matter is that the force of attraction varies directly as the product of the two attracting masses, and inversely as the square of the distance from their respective centers of gravity, and that the force is constant and produces uniform acceleration in both bodies toward each other.

Now according to that law a projectile moving at right angles to the direction of the attractive force will be constantly accelerated towards the centre of the earth, and in consequence will describe a parabola in its motion.

Whatever its horizontal velocity may be it will always fall sixteen feet in the first second, consequently it requires a velocity of nearly 25,000 feet per second to describe a concentric curve to the earth's surface. But I find that if I make a catapult projectile travel fifty feet in half a second, the projectile instead of falling four feet in that time, as it should do if it was subject to the force of attraction, it only falls a very little over eighteen inches. Now this is a crucial test of attraction.

If there is such a thing as attraction in matter, no horizontal velocity less than 25,000 feet per second will prevent the projectile from falling to the earth. Now according to my discovery, and the law of kinetic stability, a horizontal velocity of 1,036 feet per second kept constant will enable any weight of projectile from falling to the earth, and I will explain to you how kinetic stability produces that effect.

First then we know that any mass m with velocity v in a straight line uninfluenced by any other motion, having perfect freedom, has $\frac{m \, v}{2}$ as its kinetic energy. It must have had that amount of work expended on it to give it the velocity v, and it is capable of doing that amount of work before it is brought to rest.

If instead of a mass whose inertia is proportional to m we have a weight w on the surface of the earth whose acquired acceleration in one second in falling towards the center of the earth is $g = 32.2$. Then if $E_k =$ kinetic energy $= \frac{w \, v^2}{2 \, g}$. That we all know; there is no difference in our opinion on that point.

But I have discovered that the mass m or weight w has not only kinetic energy in the line of its motion but it acquires kinetic stability transverse to the line of its motion, and that its kinetic stability always varies in the same ratio as its kinetic energy varies. Now that fact is not known, nor yet accepted, in the science of mechanics or dynamics, and on this fact depends, as I will show you, the whole theory of gravitation and the true cause of what has been hitherto considered as due to the mutual attraction of matter. I am sure every one of you in your experience must have noticed that the greater the velocity given to a body it became more difficult to deflect it, and if sufficient force was not exerted it would tend to oscillate, or appear to be attracted to the line it was previously moving in. You have all noticed how a spinning top recesses when it is displaced a little, and how a gyroscope vibrates across the line of its motion; how a bicycle keeps erect in motion, and requires considerable force to make it oscillate, but which when at rest has to be supported from falling.

Those of you who have seen the water issuing from a monitor with a 500-foot head would see the difference in stability from the same water under a 50-foot head. On this principle depends the ejector and injector, and various other mechanical inventions whose mathematical theory, for want of this simple law of kinetic stability, is as intricate as the trigonometrical epicycles of Ptolemy when he proved that the sun went

round the earth, to the satisfaction of himself and
all the scientific papers of the world for 1,400
years; and when Copernicus thought the same
thing would happen if the earth simply turned
on its axis, he was thought an eccentric old fool
to try and alter the laws of Nature which God
had established from all eternity.

They even went so far as to say that he might
as well try to prove that two and two make nine
as to prove that the earth turned on its axis. It
was a terrible idea to the multitude to think that
the earth turned on its axis, and the comical
illustrations of that period are as amusing almost
as what appeared the other day in the San Francisco papers about Prof. Stevenson, who did
not believe that one piece of matter attracted
another. Well let it pass, we are content to be
like Copernicus, persecuted for the sake of
scientific truth, knowing full well that the laugh .
will turn some day.

Now if a body has kinetic stability equal to
its kinetic energy, it will vary as the square of the
velocity of the body, and in different bodies as
their masses, and that is the new law of kinetic
stability for a body moving in a straight line.

Now we will consider the motion of a body in
an orbit, which is a circle. If a body moving in
a straight line has kinetic stability equal to its

kinetic energy, then so long as the body moves in that line its kinetic stability is latent, and on this fact we can explain many of the latent energies amongst molecules, such as the latent heat of steam, and by a true understanding of this fact we can find out how to convert all their latent energies into kinetic or work-producing energies, which will be one of the earliest outcomes of the application of this discovery to mechanical and economical inventions.

Now let us investigate circular motion, where the kinetic stability has been all transformed into potential energy, and we will find that the kinetic stability is now active as Newton's centripetal force, the centripetal force is there, but it resides entirely in the moving body, not in the center to which it tends, as we will prove when we have circular motion without any mass in the center to attract the body, and we believe there are planetary systems in the universe which have no sun in the center of the orbits, and what we call twinkling stars are sunless systems whose planets are themselves suns revolving round a sunless centre, the possibility of which we will prove to you later on.

We know that the centrifugal tendency of a body constrained to move in a circle is expressed for the same body in different circles as $\dfrac{m\,v^2}{r}$

where r is the radius of the circle, and v the linear velocity in the orbit. Now if we simplify the problem by taking a circle of radius 1 we have the centrifugal force proportional to $m\ v^2$, that is, the stress on the cord fixed at the centre to retain the body in the circumference of the circle; now the actual stress due to the body is one half of that, viz: $\dfrac{m\ v^2}{2}$ because the other half of the stress is re-action on the fixture at the centre, which we can prove by replacing the fixture by another string of equal length fastened to another body of equal mass, moving with equal velocity in the same circle. The stress is now the same, although there are two bodies instead of one, but the centrifugal force for both is $m\ v^2$, therefore for the one it must be $\dfrac{m\ v^2}{2}$ and this is another fact from which many mechanical inventions will follow.

Now the centrifugal tendency due to the circular motion of the body is proportional to $\dfrac{m\ v^2}{2\ r}$ and if the radius of the circle is unity it is proportional to $\dfrac{m\ v^2}{2}$. Now the kinetic energy of the body moving in the circle at right angles to the centripetal force, which counteracts the centrifugal, is also $\dfrac{m\ v^2}{2}$, hence the centrifugal force is equal to the kinetic energy, and the kinetic stability is equal to the kinetic energy, therefore the kinetic stability is equal to the centripetal force, and as we have shown that a body in free motion has only got kinetic energy and kinetic stability, therefore kinetic stability must be the same as the centripetal force, or force of attraction.

Now we will prove that circular motion does not require any string to hold the body in the curve, and such being the case there can be no attractive force.

It is almost an impossibility to make a practical experiment in free space to show the effects of kinetic stability, because of the great velocity required to overcome gravity, so that before a body can be said to be moving with perfect freedom on this earth it must have an independent velocity of at least 1,036 feet per second.

I suggested the following experiment to Lord Kelvin last summer: To get Sir Edwin Watkin's permission to use the 1,200-foot tower now being built near London, have an iron tube about four feet diameter in the centre of the tower, so that it could be exhausted of air, have partitions of steel paper every ten feet, let them be in an electric circuit, so that a record would be made of the instant a bolt fell through each, and the distance from the centre.

Drop a soft iron bolt from the top of the tower; have an electro-magnet placed 200 feet from the top of the tower, of sufficient capacity to deflect the bolt in its passage about one half inch; now if such a bolt had a uniform velocity, and be deflected by an impulse it would in its further passage continue to oscillate across the original line of its motion, due to its kinetic stability, or the *persistence of energy*. Such an experiment would prove the elasticity of motion, and, as Lord Kelvin says, if proven would be the fulfillment of the golden dream of science.

Now this law of kinetic stability shows that motion is elastic, because if it requires a force greater than the body's kinetic energy to permanently deflect a body, then if the force applied be less than the kinetic energy the difference between that and the kinetic energy is a force of restitution, which is the same as the force of gravity, and follows the same law, being inversely proportional to the square of the distance between the centre to which the body tends and the centre of the body itself, and it is that *unearned increment of the body's kinetic stability*, which we call the force of restitution, or in the case of the planets the force of gravity, and that force varies in all conic curves inversely as the square of the radius vector. Of course in the case of the circle under uniform velocity there is said to be no force, or rather the centripetal force is balanced by the centrifugal force, but in an ellipse where the radius vector joining the focus towards which the force of restitution tends and the body, the force of restitution is inversely proportional to the square of the radius vector.

I have gone over all Newton's propositions respecting this force in the " Principia," and this force of attraction in every case is the same as the force of restitution, and his calling it an attractive force is simply a case of mistaken identity, in fact a delusion similar in every way to the Ptolemaic delusion. The centripetal force appears as if it was an attractive force, but is in reality only a force of restitution.

Energy in its most general sense is purely a mode of motion, there must be resistance coupled with velocity, and if $m =$ the resistance, then $m \, v^2 = E =$ energy, or *vis viva*, but m may be taken as small as you please, so long as it is not nothing, and v may be taken as great as you please so long as it is not infinite.

Kinetic energy $= E_k = \dfrac{m \, v^2}{2}$ is the kinetic energy due to direct force, and is a measure of the work done on a body in motion.

Potential energy $= E_p =$ the capacity of a body to do work.

Kinetic stability $= E_s = \dfrac{m\ v^2}{2}$ is the resistance offered by a body moving with independent motion to change of direction, and is always reciprocally proportional to the potential energy.

Kinetic stability is a kind of latent energy.

Force $= f = m\ a$, is the imaginary agent by which power communicates an action against a resistance, and it may truly be called the spirit of power. An effective force is an unbalanced effort.

We divide forces into two general classes and several specific varieties.

First.—*Conservative Forces* are those which increase both the kinetic stability and kinetic energy at the same time, and in the same proportion.

Second.—*Radical Forces* are those which increase the potential energy at the expense of the kinetic stability.

Specific kinds. First class centripetal force $= f_{cp}$ is generally the tendency of a body in motion to return to the original line of its motion if disturbed by a force less than the kinetic stability of the body.

Second class centrifugal force $= f_{cf}$ is the tendency of a body to move from the centre, being a component to the tangential motion of a body moving in an orbit.

Force of Restitution f_r is the excess of the conservative forces over the radical forces, and it is equal to the *unearned increment of the kinetic stability*, and is the same as Newton's force of attraction.   •

PROPOSITION I.

Every body which has independent motion has kinetic energy equal to the work done by a conservative force in the direction of its motion $=$

$$\frac{m \ v^2}{2} = E_k \ .$$

Let $m =$ the specific resistance of the body to change of motion, and v the velocity generated by the action of a conservative force in time t. Then the action of a uniform force starting from no velocity to v velocity is equal to $\frac{v_0 + v}{2}.$

Now $R \ s =$ work done, where R is uniform resistance and $s =$ distance. Now $R \ s = f \ v \ t = m \ v^2$, and the mean work when f starts from O is $\frac{m \ v^2}{2}.$

PROPOSITION II.

Every body which has kinetic energy has also kinetic stability, and its kinetic stability increases with its kinetic energy.

Let a body moving with kinetic energy $\frac{m \ v^2}{2}$

be disturbed transversely in its course, and we
know that the force necessary is directly proportional to the square of the velocity, and inversely
as the radius, now if the mass m and the radius
r be constant the force varies as the square of the
velocity.

Now we know that the kinetic energy varies as
the square of the velocity, hence the kinetic stability increases in the same ratio as the kinetic
energy, from m to $\dfrac{m\,v^2}{2}$.

FIG. 2.

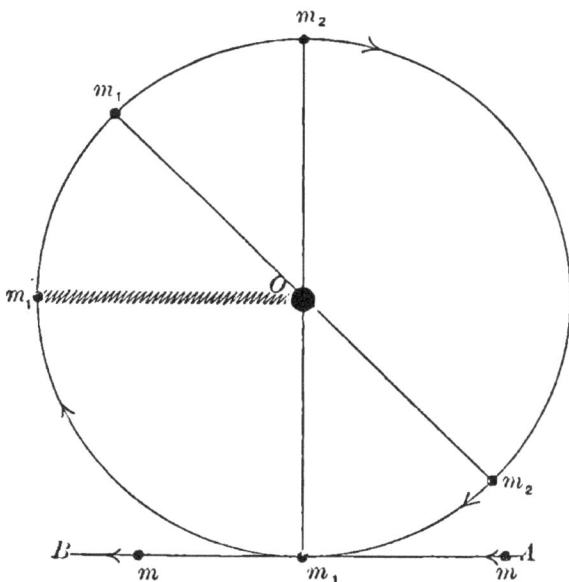

PROPOSITION III.

The kinetic stability is equal to the kinetic energy.

Let us take the body m moving with velocity v in the straight line $A\ B$. Hang another body $m_1 =$ to m from a point O by an inextensible cord, so that the body m_1 will be at rest in the straight line $A\ B$, and m will impinge and give up all its energy to m_1. See Fig. 2.

Now m_1 by means of the cord and fixture at o will be constrained to move in the circumference of a circle of which the cord is the radius, and with the same velocity in the curve that it had in the straight line, which proves that it has the same kinetic energy as formerly.

But we also become cognizant of another stress or energy, or capacity to do work along the cord, which we can measure by a spring balance inserted in place of the cord and fixed to O, and the measure of that stress as shown by the balance is $m_1\ v^2$, and in different circles is $\dfrac{m_1\ v^2}{r}.$

But the actual stress due to $m_1 = \dfrac{m\ v^2}{2}$, that is easily proven by using another mass $m_2 =$ to m_1 caused to revolve by an extension of the cord, forming a diameter to the circle. There will then be a balanced stress on the fixture, and the stress on the extended cord will be $m\ v^2$, although two equal masses m_1 and m_2 are moving in the circumference with equal velocities, and with a kinetic energy in each equal to $\dfrac{m\ v^2}{2}$, yet the stress on the cord does not alter from $m\ v^2$, although the masses and the kinetic energies have been doubled, which proves that the transverse stress due to the kinetic energy of m is not $m_1\ v^2$, but $\dfrac{m_1\ v^2}{2}$.

Now the mass m_1 in describing the complete circle impinges on m, and gives up all its energy to it, that is, supposing the bodies are perfectly elastic and under frictionless conditions. Now m continues with its original velocity in $A\ B$, but where is the transverse stress which was shown on the cord whilst describing the circle? We know it is not in m_1, which is perfectly dead, and as no energy can ever be destroyed completely, it must be in some way attendant to m in its motion, and we can prove that it accompanies m as a latent energy capable of being made potential by means of a radical force, and such quality of motion we call the body's kinetic stability, and we have shown that it is equal to the kinetic energy $= \dfrac{m\ v^2}{2}$.

PROPOSITION IV.

If a body in motion in a straight line be disturbed by a force transverse to the line of its motion, and the amount of that force measured by the energy it produces is less than the kinetic stability of the body, then the body will be acted on by a force tending to make it return to its original line of motion, and such a force I call the force of restitution, and its value is equal to the difference between the kinetic stability and the amount of the disturbing force, that is, $f_r = f_g = E_s - f_d = E_k - f_d = E_p$.

On the proof of this proposition rests the most wonderful property of energy, the *elasticity of motion*, the golden dream of all philosophy. The corner-stone of the most enduring edifice which science can ever build, whose foundations were laid in the beginning of time, and on whose glorious turrets rests the spiritual emancipation of the human soul.

In my next paper I will prove the above proposition, and show its application to mechanical invention. We will now conclude by stating the *Law of Kinetic Stability.*

A body moving in free space with independent motion has kinetic energy in the line of its motion, and kinetic stability tending to prevent displacement transverse to its line of motion equal to its

*kinetic energy, and should a radical transverse
force be applied equal to the kinetic stability, the
body will then move in a curvilinear orbit with a
velocity equal to $\sqrt{2} \times v$, v being its original
velocity, and according as the transverse radical
force is greater or less than the kinetic stability, so
the curve will be one of less or greater stability, and
the force of restitution will be less or greater.*

And the Law of the Persistence of Energy:

*Every body moving with independent free motion
in a straight line has a specific tendency to continue
in that line, varying with and always equal to its
kinetic energy, and if displaced in free space by a
transverse force there will always be an unbalanced
component force in the body's motion tending to
bring the body back to the continued line of its
previous motion.*

The unbalanced component is what Newton
ascribed to the mutual attraction of matter, but
which we now know is due to the *Persistence of
Energy*, and we call it the *Force of Restitution*.
That is the universal force of gravitation; it is
always inversely proportional to the potential
energy.

Now as the potential energy of a displaced
force varies with the square of the displacement,
therefore the *Force of Restitution*, due to the
persistence of energy, will vary inversely as the

square of the distance or displacement. That is the geometrical law, and is true for kinematical motion, but the great kinetic law depends on kinetic and centrifugal energies.

———

THE KINETIC THEORY OF GRAVITATION.

———

PART II.

In my former paper on "A New Theory of the Cause of the Centripetal Force," which Newton and his followers have ascribed to the mutual attraction of masses, I mentioned the fact that the kinetic stability of a body which has independent motion in a plane normal to the centripetal force of gravity, would tend to neutralize the apparent effect of that force, and that in a succeeding paper I would show how that result was produced.

In the discussion on the paper it was suggested by some members that the question was of sufficient importance to be submitted to the professional scientists at Berkeley, and Palo Alto.

Professor Soulé of Berkeley endorsed my statement that the experiment of the catapult properly carried out would be a crucial test. But he said in effect, that the force of the mutual attraction of masses had been a settled doctrine in science so long, and seemed to satisfy every requirement, that he would not be prepared to doubt its existence, until after a most exhaustive proof that it was due to something else.

Exception was also taken by some to the loose way in which I had put the argument, and to the want of details in the catapult experiment.

The experiment was one of many which I made fifteen years ago, when I first doubted the theory of attraction, but as these experiments were not carried out in a vacuum I did not consider them sufficiently exact to be called scientific experiments, but only approximations, and although they were sufficiently approximate to show me that a body in motion in a plane at right angles to the force of gravity did not fall with the same centripetal acceleration that the same body would do if at rest, yet they were not crucial enough to place before the Royal Society of London.

In reading up the subject in tracts, published by the late Dr. Hutton of Woolwich Academy, England, on the experiments in gunnery, I found that that eminent mathematician had noticed the same discrepancy between the fall of bodies due to attraction, and the practice with projectiles at high velocities, which he ascribed to some polarized condition of the air. (The word polarized, at that time was a fashionable fad with scientists for explaining away any discrepancy to an established theory, having been brought into prominence by the researches of Dr. Thomas Young and Dr. Brewster in connection with light.)

Nearly every scientist since, who has noted the discrepancy, has tried to explain it, by supposing the projectile to kind of hypnotize the layers of air immediately above and below it, so as to prevent it from following the law of falling bodies whilst in motion.

Instead of accepting that explanation, I always felt sure there was some other cause at work, and the result of my studies and experiments have led to the discovery of the true cause, viz: that the *persistence of energy*, and the accompanying *kinetic stability*, when subjected to lateral forces, is the true cause of gravity.

When I mentioned the matter to my old professor, Lord Kelvin, he thought kinetic stability was too intricate a mathematical function, to use in explaining such a simple matter, and to approach the subject mathematically from that hypothesis would exhaust the methods of the analytical calculus.

Consequently in these papers I prefer to go back to first principles, which even the undergraduate can understand and follow, and so build up my theory on the very simplest application of the laws of motion.

In the discussion, I was pleased to know that Mr. Richards, a past president of the Society had given the weight of his large experience in favor of the theory of kinetic stability as a probable cause of centripetal force, in preference to Newton's theory of attraction.

He finds in his practice many actions and reactions in high speed centrifugal water pumps, for which empirical formulae have to be substituted for theory, for the simple reason that the present theory of deviating forces does not meet the requirements of the case.

Now as it seems to be the almost general wish of those members who have considered the matter, to have my theory submitted to one or both of the principal universities in this State, I will therefore devote this paper to a description of the experiments, which, if carried out with scientific accuracy, will prove that my theory is the true one.

I need not call attention to the important results to science, which would follow from the acceptance of the theory of *kinetic stability* as the true proximate cause of the centripetal force of attraction throughout the universe.

The past, present and future conditions of each planet, not only physically, but vitally, together with all the actions and reactions between them and their primary, would be calculable from their motions alone.

Whether they were hollow or solid spheroids; the specific heat of the mass of each; the extent and density of their atmospheres; the specific force of gravity on their surface; the character and bodily structure of their inhabitants; would all become mathematical problems, depending for their solution on the ratio of the centripetal to the centrifugal energies of each planet.

Then when we carry the question of kinetic stability into molecular physics, it opens up to exact science problems which, when solved, will revolutionize all our ideas of matter, soul and spirit, and show us how to produce and control the vast latent forces of nature, with greater efficiency than we have hitherto dreamt of.

If these are not a sufficient encitement to a proper determination of this scientific subject, then the love of truth itself, which is strong in every true scientist, should be motive sufficient for the purpose.

Let us now consider proposition IV.

If a body in motion in a straight line, having kinetic energy $= \dfrac{m\,v^2}{2}$, be disturbed by a force acting transversely to the line of its motion; if the disturbing force be *any finite force whatever*, the disturbed body will describe a curve as the *resultant motion*, being acted on by two effective components, the one towards the instantaneous centre of curvature, we call the *force of restitution* or centripetal force; the other acting normally to the centripetal force, we call the *tangential force*.

This proposition has never been accepted hitherto by dynamical science, nor to my knowledge has it ever been completely demonstrated.

If the simple action of two impressed forces acting at right angles, will cause a body to describe a curve as the resultant motion, then where is the need for the mutual attraction of masses to cause curvature?

I need not refer to the established principle in dynamical science, called the resolution of forces or motions, which proves that the resultant of two forces or motions, is the diagonal of the parallelogram whose adjacent sides represent the forces or motions in amount and direction.

That I acknowledge as a cardinal principle of the science of dynamics, but it only applies to cases of constrained motion, and I am dealing with *independent motion in free space.*

To illustrate my meaning; suppose a ball be rolled across the platform of a car in motion, then the resultant motion, both in direction and amount, would be expressed by the length and direction of the diagonal of the parallelogram, whose adjacent sides represented the direction and amount of the respective motions.

But suppose we give the ball a certain finite velocity along the car, independent of the car's motion, and also give it a transverse velocity across the car, then although the diagonal of the parallelogram as above would give the velocity due to the constrained motion or momentum, yet the resultant of the independent motion would be a curve.

I can demonstrate that fact with mathematical accuracy, and can show the value of the coördinates of the curve under various forces, and one remarkable result of the geometry of the subject is, that *however small the kinetic energy of the body may be, so long as it is finite; and however great the transverse deviating force may be, if it is not infinite, the body can never be permanently displaced from the original line of motion, but will always have a component force acting as Newton's centripetal force apparently towards a focus.*

And that is a most wonderful fact which explains the constitution of the universe as a stable or conservative system, which no radical force less than that of Omnipotence can ever destroy.

And it also teaches us another most wonderful principle in the constitution of the universe, viz: that the forces necessary to give it stability, must have been controlled and directed by some power independent of and unaffected by the phenomena produced.

That in fact the Creator of the physical universe, must have done work to produce the phenomena which we call the universe. It is that principle of independent action, which acting against kinetic stability makes stable motion.

Again let me explain my meaning. If a body like a railway train, or a cannon ball, be constrained to move in a curve, by means of a frictionless guide, then no additional energy has been communicated from without, but simply the resistance of a frictionless guide; hence no energy has been added or extracted from the moving body. The orbit has been described by means of the inherent energy, or *vis viva* of the body, and the result is that the moment the guide is removed, be it a string fixed at a centre, or an attractive force from a centre, or a rigid friction-

less guide, then the moment that it is removed
the body goes off on a tangent to the curve it has
been constrained to move in.

But suppose the cannon ball receive an impulse
from an independent outside force, acting transversely to the line of motion of the cannon ball,
equal to twice the kinetic energy of the ball,
then the ball will forever move in an orbit which
is curvilinear, and if the deviating force exceed
twice the kinetic energy, the ball will move in
an orbit which is an ellipse of any eccentricity
whose centripetal force will vary inversely as the
square of the distance from one of the foci, and
the resultant motion is perfectly stable, the same
as the universe, or the solar system.

As the mathematical proof of this proposition
is purely theoretical, I will not complicate this
paper by introducing it here, but will proceed to
sketch out a method of experimental demonstration, which will establish the fact that curvilinear motion is the true resultant in all free
independent motions.

EXPERIMENT I.

Take a slab of polished marble or glass, or any
other polished surface, perfectly horizontal, or
level in every direction on the surface, about
fifty feet long, and ten feet broad. At one end
fix a catapult arrangement, which will project

a spherical steel ball (of the size of a billiard ball), along the top of this slab in a straight line longitudinally with a velocity of one hundred feet per second.

About one foot from the initial point of motion of the longitudinal catapult, place a transverse catapult, having its initial point of motion one foot from the line of the other. Let both catapults be capable of giving the same velocity to the same sized balls of steel. Let the springs of both catapults be held back by the one string, the centre of which is placed between the carbon points of an electric machine, which by burning the string allowed both catapults to act instantaneously. Then, if proper precautions are taken, the transverse ball will strike the longitudinal ball one foot from its initial point, with the same kinetic energy as the longitudinal ball has.

The resultant motion of the longitudinal ball will be curvilinear, and whatever energy the transverse ball loses, the other will receive, less the friction of course and the viscosity of the balls, and the various energies of heat, light and electricity, dissipated by the concussion. The curve can be shown by having a black slab and a whitened ball.

By careful experiment in a vacuum, with various degrees of energy, the coördinates of various curves can by this means be almost exactly determined, and so one of the most important principles in the whole realm of dynamical science can be demonstrated by experiment. And I have no hesitation in saying, that the professor, and the university, which demonstrates the truth of this principle most correctly, will earn for themselves a name and a fame, which will be esteemed and honored whilst the world lasts.

To such an institution the ambitious youths from every nation will come for instruction, because the establishment of this principle of kinetic stability, is one of more far reaching importance to science in general, than we can imagine at present, as not only physical phenomena, but vital and spiritual phenomena all depend on it largely for true scientific treatment. In this great principle resides the hidden power which through the method of evolution does its work. And on this latent energy depends the influence which enables faith to control the phenomena of the spiritual world.

The theoretical proof will be given by me in another paper, as soon as the experimental proof is ready for publication; to show what is the difference between theory and experiment. A knowledge of the theory tends to make the experimenter imaginative, and unreliable in the statement of results, hence the theoretical demonstration will be witheld for the present.

The second problem is to prove experimentally, that a body near the surface of the earth moving horizontally with kinetic energy, will not be influenced by the force of gravity with the same acceleration towards the centre of the earth, as if the same body had been at rest relative to the earth. Or in other words, a horizontal projectile moving one hundred feet in one second, will not fall sixteen feet during that second.   .

To prove this by experiment exactly, and find the law of its acceleration under various horizontal velocities, requires a very complete and elaborate apparatus, such as only a university can provide, but this would be a crucial test of the two theories:

Kinetic stability versus *attraction of matter*, and as I know that kinetic stability is the true one, and is bound in the near future to work a great revolution in all science, it is well worthy the earnest attention of all teachers of science, and those who wished to be honored as the leaders of the scientific world, will do well to look to their laurels.

Here is a simple experiment, which if carried out carefully *in a vacuum* will upset the best-established theory in all the domain of natural science, which will cause half of the scientific books in the world to be re-written.

Can we rest satisfied, and continue to teach fallacies to the world, and dignify them with the name of science, when such a simple experiment will forever correct and remove this error?

Although this proposition can also be demonstrated with mathematical accuracy by using the principle of kinetic stability, yet the three laws of motion will not enable us to demonstrate it, hence it is necessary that an experiment of a crucial character should establish the fact before the scientific world will believe in it.

Now the question is, to which university shall the honor belong in having with the strictest scientific accuracy demonstrated to the world this important fact; viz: that a body moving horizontally with independent motion will not fall sixteen feet in one second, the distance it would fall if at rest?

Although I would prefer to leave to the experimenter the choice of methods to determine the truth, yet there are certain precautions to be taken which it may be well for me to mention.

1st. The experiment must be carried out in a vacuum, as explanations have already been attempted, to prove that the air carries the body, and so prevents the fall.

2nd. The velocity should exceed fifty feet per second.

3rd. The apparatus should be as rigid and unaffected by vibration as possible.

4th. The air-exhausted tube or case should be at least fifty feet long with a properly-stretched diaphragm mounted on a square frame placed at distances apart of one foot; two transverse lines cutting each other at the center of the diaphragm, as impressions or water marks in the diaphragms should be ranged exactly by means of a transit instrument. the diaphragms should be white and the projectile blackened.

5th. The vacuous box should have one side removable with great ease to enable the trajectory to be exactly measured.

6th. The time should be if possible one second.

As the value of this experiment to mechanical and astronomical science cannot be overestimated, it behooves the experimenter to use the greatest care in devising and describing each and every detail.

As aerial locomotion is bound to depend very largely on the fact that a body in motion does not fall the prescribed sixteen feet in a second, it can easily be seen of what immense importance the knowledge of this fact will be to the mechanical and political world.

Important as that fact is to mechanics, it is no less so to astronomy, as the law of attraction depends at present on the multiplication of masses. If 10 units of mass be multiplied by 20 units of mass the resultant attraction is at present equal to 200 units of mass, but we know that the quantity of matter contained in two bodies, one of which contains 10 units of matter, and the other 20 units of matter, can never be more than 30 units of matter.

Now although the law of the inverse square of the distance enables astronomers to make close approximation in the motions of the planets, the law of the multiplication of the masses enables them to make ignorant guesses at the specific gravity of each planet, and the specific heat which the mass contains, and by this means the world is kept in complete ignorance of the conditions of the inhabitants in the other planets.

Now when we look at this subject from every possible point, it behooves us as a Society, either to act alone, or in conjunction with the Astronomical Society, and use all the influence possible to have this matter correctly determined by one or both of our great universities.

It behooves us also to act promptly, as the matter will be taken up by other institutions in the Eastern States, who, from the many letters I receive, seem to be acting in earnest, and appear to take more interest in the subject than the scientists of this Coast.

That California should have the high honor of demonstrating to the world the principles of kinetic stability, and proving that the force of the mutual attraction of masses is simply a fallacy and a delusion, is the most sincere wish of a humble member of the Technical Society.

www.ingramcontent.com/pod-product-compliance
Lightning Source LLC
Chambersburg PA
CBHW021515090426
42739CB00007B/623